For Tsuguko, Noriko, and Ayako

Oxford University Press, Walton Street, Oxford OX2 6DP

Oxford   New York
Athens   Auckland   Bangkok   Bogota   Bombay
Buenos Aires   Calcutta   Cape Town   Dar es Salaam   Delhi
Florence   Hong Kong   Istanbul   Karachi
Kuala Lumpur   Madras   Madrid   Melbourne
Mexico City   Nairobi   Paris   Singapore
Taipei   Tokyo   Toronto

and associated companies in
Berlin   Ibadan

Oxford is a trade mark of Oxford University Press

Text © Her Imperial Highness Princess Takamado 1996
Illustrations © Brian Wildsmith 1996

First published 1996

This book has been produced in association
with the Brian Wildsmith Museum of Art

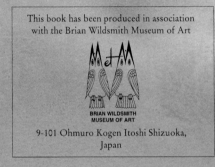

BRIAN WILDSMITH
MUSEUM OF ART

9-101 Ohmuro Kogen Itoshi Shizuoka,
Japan

A CIP catalogue record for this book is available
from the British Library

ISBN 0 19 279005 6

Printed in Hong Kong

Katie *and the* Dream-Eater

Her Imperial Highness Princess Takamado
Illustrated by Brian Wildsmith

Oxford University Press

Far, far away, beyond the rainbow and above the clouds, there is a land where all the fantastic creatures live—unicorns, dragons, and many, many others. Here, too, live the baku, the shy creatures that eat up bad dreams. Every night the baku go forth into the world in great numbers to make sure that everyone has a good night's sleep.

One day, much like any other, a baby baku sat waiting for his mother and father to come home. Fighting and eating nightmares is very dangerous and so they had to leave him behind. But he was a brave little baku and he wished he could go out and fight bad dreams, too.

'I will go and see what is happening in the world of people,' he thought. And away he flew. He saw many people fast asleep with smiles on their faces, but he also saw a little girl, called Katie, tossing and turning in her bed.

She looked like a kind and gentle girl, and he was sorry to see her so frightened. None of the other baku were near, and he knew that it was up to him to fight her bad dream. He took a deep breath and charged.

Katie was terrified by her bad dream. As she ran away, she could feel its breath. It was just about to catch her—when suddenly she saw a baku in her dream. The brave little baku attacked the bad dream, but the dream monster was very big. Katie saw it raise its head and prepare to strike.

'No!' she shouted. 'Leave him alone!'

Katie threw herself at the monster. She tweaked its nose. Then she started to tickle it. The monster began to giggle and it turned bright red. No one had ever tickled it before. It wriggled and squirmed and roared with laughter.

'Quick!' said Katie to the baku. 'We must run away before it stops laughing.'

Then . . . Katie woke up. She picked up her favourite teddy
and gave it a hug. She was very tired but happy to be in her
own bed. She looked around her room.

'Oh!' she gasped, and her eyes opened wide. The baby baku
was standing beside her, sucking his nose. 'You can't be here!
You should be on the other side of my dream!'

Morning came. Katie introduced the little baku to her mother and father and then took him to school. There, the teacher prepared a special desk for him and showed him how to use his books. When it was time to play, her friends showed him how to ride on the swings and how to see-saw. Together they watched a naughty boy, called Peter, chase a cat up a tree and throw stones at the birds.

That first day, and for the next few days, the little baku had great fun. But he was hungry. Katie knew that she had to feed him with bad dreams, but she was now a very happy girl. She no longer had nightmares.

'We must go out and find you something to eat,' she said.

So that night, the two of them slipped quietly out of the house in search of bad dreams.

The little baku looked through many windows.
He rescued a cat who was being threatened by a giant eagle.

He left Peter alone. 'Perhaps his dream will teach him a lesson,' he said.

He helped a little boy who was having a bad dream about all his sums.

And finally, he helped a dog whose bone had got caught up in a giant kite.

The next day, Katie had an idea. 'I'll take you to the hospital today,' she said. 'There are lots of sad children in there. Perhaps you can help them.'

So after school they went to the hospital. Katie introduced the little baku to the doctors and nurses.

Then she said to him, 'I'll wait for you at home. When you've had enough to eat, please hurry back.'

But many days passed, many weeks passed, and the baku did not come home.

Katie was sad and lonely. One cold and rainy day, she walked to the hospital to ask him to come back. But then she saw how everyone needed him. 'I mustn't be selfish,' she thought.

In the evening, Katie was taken ill with a fever. The doctor told her parents that she should be in hospital.

That night, the doctor brought the little baku to her bedside. Every time she had a bad dream, the baku ate it up. When she opened her eyes, he spoke of happy things and the wonderful creatures in faraway lands.

Then one day Katie had a very bad dream. The little baku tried to eat it, but the ferocious dream monster attacked him. The brave little baku threw himself at the beast but he wasn't strong enough.

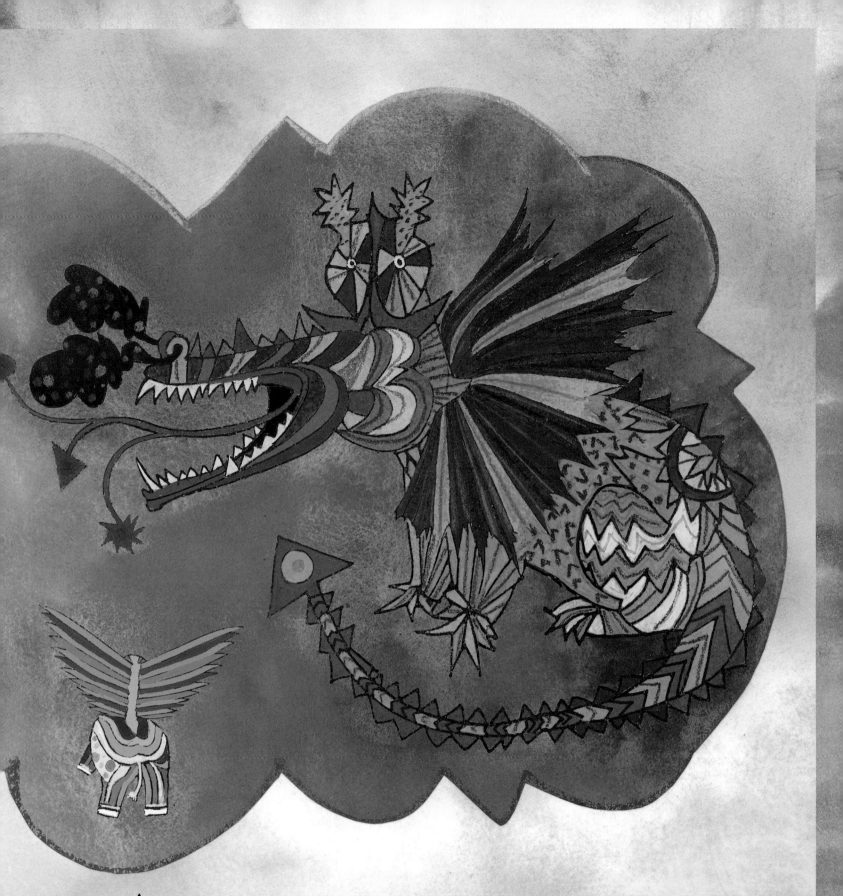

As she slept, Katie saw the baku being chased by her nightmare. 'Someone help him!' she cried. 'Please help him!'

Then suddenly two big baku appeared out of the clouds and swallowed up the terrible monster in a flash. The baby baku cuddled up to his mother. The father baku thanked Katie for looking after their son. As they turned to go away, the little baku ran back to her, whispered something, and then hurried after his parents.

Katie woke up. Her eyes were shining. 'He's coming back!' she said. 'He says he'll bring some unicorns and dragons and others!'

'What a lucky girl you are!' said her mother. 'I hope the house is big enough for all your new friends!'

'No, no!' said Katie. 'Next time we'll play on his side of my dream. I don't need to have a nightmare to see him. Isn't it wonderful?'

My friend the Baku
by Katie